D1406749

A65500 167403

TRACTORS

From Yesterday's Steam Wagons to Today's Turbocharged Giants

by Jim Murphy

J.B. LIPPINCOTT   NEW YORK

For Sean Michael Murphy

Acknowledgments

I would like to thank the following individuals and organizations for their valuable assistance in assembling information and photographs: Les Stegh, Archivist, John Deere & Company; Robin M. Bird, Manager of Marketing Communications, J.I. Case; Dr. Lee Phillips, Assistant Curator, F. Hal Higgins Library of Agricultural Technology, University of California; R. G. Walther, Museum Specialist, Smithsonian Institution; L. Y. Johnston, Media Placement Supervisor, Bolens Corporation; GemPhoto; Debbie Scalara and Cindy Arruda, Gemini Graphix.

Illustration credits: Bolens Corporation, 52; J. I. Case, 30, 31, 32, 36, 37, 42, 43, 46, 47 (bottom); John Deere & Company, 39, 42, 47 (top), 48, 49, 51, 53, 54, 56; GemPhoto, title page, 7, 8, 10, 11, 13, 15, 17, 18, 19, 22, 23, 24, 25, 26, 27, 28; F. Hal Higgins Library of Agricultural Technology, University of California, iv, 34, 35, 38; Smithsonian Institution, 5, 40, 44.

Tractors
Copyright © 1984 by Jim Murphy
All rights reserved.
Designed by Trish Parcell
1 2 3 4 5 6 7 8 9 10
First Edition

Library of Congress Cataloging in Publication Data

Murphy, Jim, 1947–
    Tractors.

    Bibliography: p.
    Summary: Relates the history of the tractor and shows the changes in design that have resulted in the diesel-powered giants of today.
    1. Farm tractors—History—Juvenile literature.
[1. Tractors—History]   I. Title.
S711.M78 1984   629.2'292   82-48777
ISBN 0-397-32050-7
ISBN 0-397-32051-5 (lib. bdg.)

Contents

Seven-foot-tall metal wheels

Introduction

Every morning farmers around the world climb aboard their tractors to begin another day's work. Some will use their machines to plow long, deep furrows. Others will pull farm equipment or haul tons of food to market with them. There's little doubt that a tractor makes the hard work of the farmer easier and faster to do.

But if we travel back through the years, the shape of the tractor changes. Wooden and metal contraptions hissed steam and belched smoke as they chugged along. Twenty-ton monsters with iron wheels seven feet tall cut endless rows through the crusty Midwest prairie. These machines weren't even called tractors then; they were known as traction engines, or road locomotives, or simply steamers. This book looks back over the tractor's long history and shows the changes in design that resulted in today's diesel-powered giants. All you have to do is climb aboard and turn the page.

1 Cugnot's Fire Carriage

In the spring of 1765 a strange contraption was pushed from a Paris workshop. It had three large wooden wheels and a frame, or chassis, made of heavy timbers. A massive steam engine hung from the front wheel.

The machine had been designed by a French army engineer to pull a small cannon and to carry four soldiers. The engineer's name was Nicolas Joseph Cugnot.

While Cugnot watched, one of his assistants built a coal fire in the bottom of the boiler. As the water stored above the fire heated, then boiled, steam was produced. After letting steam build, Cugnot climbed aboard and pulled the release lever.

The lever opened a valve and steam rushed through a copper hose to two giant cylinders above the front wheel. Each cylinder contained a piston that was attached to the wheel below by a long metal rod. Steam entered one cylinder and forced the piston and rod down. Very slowly, the wheel began to turn and the machine inched forward.

When that piston reached the bottom of its cylinder, steam entered the other cylinder and the process was repeated. The constant up-and-down motion of the pistons and rods kept the wheel turning.

As the machine gained speed, it began bouncing wildly along the uneven cobblestone street. Cugnot fought to hang on to the steering handle and to keep his vehicle on the road. A few minutes after starting, the machine hit top speed: six miles per hour!

People on the street must have been amazed and maybe a little frightened by what they saw. A wooden-and-metal monster was rumbling toward them, spitting out smoke and steam and making a horrible clatter on the cobblestones. On top, a man seemed to be struggling to control the beast. Never had they seen anything like this. Then, almost as quickly as it had begun, the machine ran out of steam and stopped dead. Even so, Cugnot's experiment had been a success. He had become the first person to ride a self-propelled vehicle.

Those who saw Cugnot's history-making ride began referring to his invention as a "fire carriage." They thought the fire in the boiler was what made it move. Cugnot simply called it a "steam wagon." But since it had been designed to pull heavy loads, it was actually the first tractor.

Cugnot tested his steam wagon for five years. He used lead and rags to patch the leaks in the hose and cylinders, hoping to save steam and make the engine run longer. Unfortunately, he never really got it to work very well. The boiler was small and could only make enough steam for a slow fifteen-minute ride. What's more, the machine was hard to steer and even harder to stop. In fact, one day Cugnot lost control of his steam wagon and crashed through a courtyard wall, making him the first person to have a traffic accident.

Cugnot built a second steam wagon in 1771, but he never got a chance to run it. Advisors to King Louis XV became jealous over the attention Cugnot was getting and put a stop to his experimenting.

Although Cugnot's steam wagon was never again seen rolling along the cobblestone streets, he had proved his point: steam could be used to move large vehicles and to pull heavy loads. And Cugnot's simple piston-rod setup would become the method of power for almost every self-propelled machine invented over the next 200 years.

Cugnot's steam wagon crashes into a courtyard wall. The inventor is struggling to steer the machine out of danger while his assistant tugs at the brake.

2 Experimenting with Steam

Word of Cugnot's invention spread around the world. Soon other inventors began looking for ways to improve on his work. Not many of them were particularly interested in building a tractorlike vehicle, however. Most just wanted to make a self-propelled machine that would replace the horse and carriage as the basic means of transportation.

One of the first tinkerers to successfully improve Cugnot's invention was an Englishman named Robert Murdock. Murdock was an assistant to the famous inventor James Watt. During the day, Murdock helped Watt on a variety of projects relating to steam engines. At night, Murdock worked on his own project. In 1784, he took the small, three-wheeled, one-passenger runabout he had built for a test ride.

Murdock's machine had a light wooden chassis. His steam engine was small and had only one cylinder. Instead of hooking the piston rod directly to the wheel as Cugnot had, Murdock attached it to a hinged bar overhead. The up-and-down motion of the piston rod made the bar go up and down, too. A long metal rod near the cylinder turned the two rear wheels. The driver sat under the bar, his legs straddling the center frame with a small rudder to turn the front wheel.

Murdock's runabout doesn't look very graceful, but it seems to have worked pretty well. One dark night, it whizzed through the streets of Redruth at almost twenty miles per hour. It also caused a bit of a commotion when the village parson mistook the fast-approaching cloud of steam and sparks for the devil himself.

Murdock would have perfected his vehicle. However, Watt grew upset over his assistant's experiments. Watt wanted to improve steam engines, but he simply didn't like steam-powered vehicles. When Murdock's contract was renewed, Watt added a clause that read: "no steam carriage should on any pretense be allowed to approach the house."

Robert Murdock's three-wheeled runabout was fast enough to scare the village parson and make Murdock's boss jealous.

Robert Trevithick at the controls of his giant steam carriage

The Orukter Amphibolus steaming across a rough field

It took twenty years before another Englishman, Robert Trevithick, successfully mounted a steam engine in a wooden carriage and took eight friends for a Christmas Eve ride in 1804. One witness to the ride said the carriage went up a hill "faster than a man could walk." A few years later, Trevithick put a similar machine on metal tracks, thus creating the very first railroad.

The oddest offshoot of Cugnot's work was made by an American mechanic from Philadelphia in 1805. Oliver Evans built a twenty-ton steam dredger to clear river bottoms of silt deposits. His machine came with mechanical buckets, digging devices and a paddle. The silly-looking dredger had an even sillier name: the Orukter Amphibolus.

The Orukter Amphibolus was really a combination steam carriage and boat. As in Murdock's runabout, a piston and rod at the center of Evans's engine pushed a long bar up and down. At the right of the bar, a metal rod was attached to a wheel on the deck. This wheel was known as the flywheel.

The up-and-down motion of the pole made the flywheel turn. A sturdy cloth belt attached to the flywheel made the two side wheels go around. A flip of the lever allowed Evans to drive through the

Philadelphia streets and into the Delaware River. Another lever sent the paddle into motion so he could get to wherever on the river he'd been hired to dredge.

The self-propelled machines made by Murdock, Trevithick and Evans all worked. The problem was that these inventors never tried to adapt their machines for everyday uses, such as pulling wagons or operating farm machinery. Without practical uses, their inventions remained curiosities.

Oddly enough, it was James Watt who made the next advances in steam-powered vehicles possible. Over the years Watt had made many important modifications in steam engines. He may have even read about and used some of the ideas other inventors had had about steam power. Nonetheless, what eventually evolved was known as the Watt steam engine.

Watt's engine had tighter fittings and joints, so that less steam escaped accidentally. More important, Watt found a way to recycle steam. After it pushed the piston, earlier engines let the steam escape from the cylinder into the air. Watt's engine had a small chamber attached to the cylinder. The used steam was vented into this chamber, cooled and turned back into water. The water was then returned to the boiler to be reheated into steam. This process let the engine run much longer without the need for taking on more water.

But Watt disliked the idea of steam vehicles so much that he wouldn't sell his engines to anyone making them. He sold his engines only to businessmen who used them to power machines in their factories. But by 1819, when Watt died, the patents he had held on his steam engines had run out. Inventors could use his engine any way they wanted. Soon carriages were fitted with Watt engines, and a steam-carriage boom began in England.

These steam carriages were gigantic wooden vehicles, big as modern buses, and designed to carry as many as twenty paying passengers. Two men were needed to operate these giants—one to care for the massive boiler and the other to steer.

The David Gordon steam carriage. The passengers on top look distressed, perhaps because they're sitting just above a very hot boiler.

Fleets of steam carriages were soon scurrying between cities, some hitting the then incredible speed of thirty miles per hour. David Gordon's 1824 model even came equipped with six metal legs and feet to ensure a speedy uphill trip.

But the best use of the new steam technology was made by designers of railroads in both the United States and England. First, they improved and enlarged the boiler. Next they laid the boiler on its side to streamline the engine's shape and to allow the engineer to see what was in front of him more easily. An American locomotive, the 1831 DeWitt Clinton, shows this clearly. Its engine and coal tender alone weighed about 13,000 pounds without the stagecoach-like passenger cars. Despite its weight, the DeWitt Clinton managed to chug along at around twenty miles per hour.

Only a few designers even tried to build machines to do farm work. An Englishman, Nathan Gough, manufactured one in 1830.

The 1831 DeWitt Clinton locomotive. In time, tractorlike machines would copy the general design of railroad locomotives.

Gough's machine had an old-fashioned vertical boiler with two cylinders next to it on the right. The pistons and rods turned a shaft that was bolted to the side of the boiler and extended to the rear of the machine. The illustration shows the engine set up to pump water, though it could also saw wood, drive a small grindstone or a threshing machine.

Farmers weren't particularly impressed by Gough's machine. It took over an hour to build up steam in the big boiler, and most farmers didn't want to waste that much time playing with the fire or watching valves. Worse, Gough discovered that if he tried to

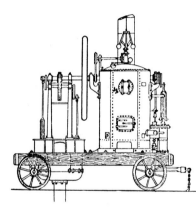

Nathan Gough's stationary engine. The short chain at the right was where the horse team was hitched.

make his machine self-propelled, almost all the steam was used up during the journey to the field. This meant even more time was needed to build up additional steam. Gough's solution was to have a team of horses pull the engine from one place to another, though this also proved a chore. As one farmer put it, Gough's engine was "about as portable as a parish church."

The idea of a self-propelled steam engine for pulling or operating equipment might have ended right here except for the vision of one Englishman, Robert Ransome, and his company. Ransome took the improvements made in steam engines, added a few of his own, and created tractorlike machines that worked.

3 Ransome's Farmer's Engine

Robert Ransome was born in 1753 and raised in the factory town of Norfolk, in eastern England. When he was thirty-three years old, Ransome was granted a patent for making the cutting blade of plows from hardened cast iron, and began manufacturing them soon after. Four years later, in 1785, he opened a second factory to produce agricultural tools.

Ransome might have been just another successful manufacturer if it hadn't been for his keen imagination and shrewd business sense. A careful study of his sales ledger showed him that new, well-made products that met the needs of buyers often resulted in a handsome profit. Soon Ransome had his company installing gas lines and gas meters, building cast-iron bridges and even making lawn mowers.

When steam carriages began crowding the highways, Ransome began thinking of ways to make a self-propelled steam engine pull and operate farm equipment. The machine would have to be powerful enough to haul heavy wagons. And it would have to be able to work ten or more hours in a row without stopping. Ransome had plans drawn up for such a machine, but he died in 1830, and his sons took over the company.

The name of his firm, Robert Ransome and Company, changed slightly after his death, becoming J. R. and A. Ransome, but the company still retained Robert Ransome's philosophy of imaginative

Ransome's 1841 self-propelled steam engine. The drawing shows the threshing device loaded on the engine and in place, with flywheel belt attached.

manufacturing. After eleven years of careful experimentation, Ransome's first self-propelled steam engine appeared.

As with Gough's engine, Ransome's had a vertical boiler. But instead of a single drive shaft to run a pumping system, the Ransome engine had a large flywheel mounted above the cylinder. A belt went from the flywheel crankshaft to the rear wheel and moved the engine forward at around six miles per hour.

Once in position, the drive belt could be disengaged to stop the machine. Then a larger belt was slipped around the flywheel to operate a thresher. To make the engine and thresher more compact and easier to move, the engine was designed to carry the thresher.

The machine was self-propelled, but a horse was always hitched to it. Why? This particular machine had no steering system. The horse was needed to ensure accurate maneuvering on the road or in the field. And even after a steering mechanism was added to steam engines, many machines still required a horse for precise turning.

The first Ransome engine was awarded a prize of £30 (about $150) at the English Agricultural Society's 1842 Royal Show. Despite the positive response, farmers wanted nothing to do with it. The reason was simple: farmers distrusted all steam machines. It seemed that every week brought lurid newspaper reports about railroad engines exploding and passengers being killed. Farmers also feared that the machine's noise would scare livestock and that the sparks would set

13

their crops on fire. And farmhands felt the new machines would put them out of work.

So strong were the antisteam feelings in the first half of the nineteenth century that several steam carriages were attacked by mobs. One account tells how "the crowd, being mainly composed of agricultural laborers, considered all machinery directly injurious to their interests, and with a cry of 'Down with all machinery,' they set upon the carriage and its occupants, seriously injuring Mr. Gurney [the driver] and his assistant engineer. . . ."

So the first Ransome engine was dismantled. The thresher was sold to a farmer, the engine to a mill owner to operate a grindstone.

Nevertheless, the people at Ransome's were sure that if farmers gave steam a chance they would see how helpful it could be to them. The designers went back to work to reduce the noise and sparks and solve other problems. After seven long years, they introduced the "farmer's engine" in 1849.

The farmer's engine resembled a railroad locomotive in design, especially because of the horizontal position of the boiler and its all-metal construction. The driver stood on a small rear platform, where he tended the coal fire and checked the water level and steam pressure. Coal was pulled along behind the engine in a tiny two-wheeled tender. A major improvement was the addition of a steering system. The steering handle was connected to the front axle by a series of gears and shafts that ran along the top and front of the boiler. Cranking the handle to the right or left made the axle turn slowly.

The farmer's engine weighed almost two and a half tons, but on level ground it could rumble along at a healthy clip of twelve miles per hour. Power to move the engine came from cylinders just above the front axle. The pistons' back-and-forth motion was transferred to the four-foot-tall rear wheels by long connecting rods and gears. Once the engine was in the field, the back of it was jacked up, and the rear wheels acted as flywheels to operate farm machinery.

"Such is the general detail of this 'farmer's engine,' " noted *The*

Practical Mechanic Journal, "which, we are sure, only requires to be pretty universally introduced. For large farms, we conceive that this engine will be a most effective assistant."

The new Ransome engine won several awards at agricultural fairs and shows. In addition to awarding prizes, agricultural societies also rated steam engines according to their horsepower. A 5-horsepower engine could pull the same load as a team of five horses. Knowing an engine's horsepower let farmers know how many horses the engine would replace.

But sales of the farmer's engine were as disappointing as those of the first Ransome engine. Farmers still preferred their reliable horse teams for plowing and harvesting. And the farmer's engine did have its share of problems. Its weight and thin wheels made it sink into soft ground easily. And while it was very well made, the pounding it took on rough country roads made gaskets and hoses shake loose.

Discouraged by the lack of sales, the Ransome company dropped the idea of a self-propelled farm engine for a while and went on to explore other types of machinery. But the seed had been sown. Robert Ransome's enthusiasm for practical experimentation had produced two self-propelled engines that could operate farm machinery effectively. And the design of the farmer's engine set the pattern that would be copied for almost seventy years.

Ransome's farmer's engine. Its wheels look thin and frail, but otherwise this machine was remarkably similar to a railroad locomotive.

4 John Fowler's Amazing Plowing Engine

John Fowler had his back to Hainault Forest in Essex County, England, as he surveyed the scene. Before him a gently sloping field stretched for 150 yards, ending at a tiny stream. On the bank of the stream, a stationary steam engine, like the one made by Nathan Gough, chugged and belched smoke while its engineer awaited John's signal.

Fowler settled himself in the seat of his invention, the mole drainer. It consisted of four small wheels attached to a simple wooden frame. Another piece of wood was fastened to the bottom of the frame and extended two feet into the ground.

Fowler waved, and the engineer pulled the lever that sent the engine into action. Instantly, a cable going from the engine to the front of the mole drainer snapped taut. The mole drainer shot forward.

Fowler struggled with the steering wheel for the first few yards, but he soon had it under control. The piece of wood under the machine cut easily through the heavy clay of the field. More important, a torpedo-shaped piece of metal at the end of the wood pushed the clay aside and formed an underground tunnel.

It took five minutes for the mole drainer to reach the steam engine. When Fowler moved his machine aside, he saw that a trickle of water was already flowing through the little tunnel. His invention

John Fowler's mole drainer would lead to a revolution in plow design.

had worked. During the week that followed, Fowler cut a series of tunnels from the forest to the stream. Soon, the water that had made the forest sodden and damp for centuries was being drained away. Three months later, it was dry enough to be cleared and planted.

John Fowler was only twenty-four years old when he designed his mole drainer in 1850, but this genius of agricultural machines had come along at just the right time. England was experiencing a population explosion. In order to feed the growing number of people, new fields had to be carved out of previously unusable land in England, Scotland and Ireland. The mole drainer did the trick handily.

Fowler didn't stop there, however. In just a few years he came up with over ninety devices to make farm work less time-consuming, including improvements in sowing and reaping machines. His most valuable invention made plowing easier.

Fowler had been impressed by the ease with which the mole drainer cut through hard-packed earth. Why not, he reasoned, use the same principle to plow a field? The result was Fowler's balance plow, produced in 1856.

The balance plow was really two plows facing each other. A seat and steering wheel on each side allowed the operator to control the machine no matter in what direction it happened to be pointed. A flagger, perched on the back of the plow, waved a red flag to signal the engineer to start and stop the steam engine.

As with the mole drainer, a steam engine and cable pulled the 250-pound plow across the field. To save time, Fowler also invented

An 1866 newspaper drawing of the Fowler balance plow

a special pulley anchor so that the plow could be pulled back and forth across the field. Otherwise, the balance plow would have had to be dragged back across the field after every run.

The anchor was positioned directly across the field from the engine and weighted with stones to keep it in place. The cable ran from the engine to the anchor and then back to the plow. As the cable was wound in, the plow shot across the field, cutting four deep furrows and giving the operator and flagger a wild, bumpy ride. When they reached the anchor, the engine was stopped and the plow tipped so that the operator and flagger could face the engine. Then the plow would be hauled back to the engine. Once the circuit was completed, the engine, anchor and plow were moved and the process repeated over and over again until the field was done.

This might seem like an awful lot of work, but it was a big improvement over the plowing methods used back then. A farmer with a horse-drawn plow could cover about one and a half acres in a ten-hour day. A steam engine and balance plow could turn three acres in the same amount of time.

Even though his plow had doubled the amount of work a farmer could do in a day, Fowler wasn't happy. Time was wasted in moving the steam engine, plow and anchor. In addition, men had to be hired to tend the horse and push the anchor along. Finally, the engines then being produced were often poorly made and in need of constant repair. Fowler knew that if his balance plow was going to be accepted by farmers, he would have to produce a complete package of steam engine, balance plow and anchor that was both reliable and easy to use.

Fowler's first step was to hook a cable to the anchor and use the power from the flywheel to move it automatically. The engineer simply pulled a lever and the anchor began crawling to the new location. Then, after five years of work, Fowler introduced his famous "winding-drum plowing engine."

Fowler's plowing engine had two cylinders on top of the boiler just behind the smokestack. The rods went to the rear of the boiler and turned a thick metal crankshaft. The crankshaft made the large flywheel turn. With the pull of a lever on the left side, the crankshaft power could be transferred through a beveled gear and vertical shaft to turn the winding drum on the belly of the machine. On the other

The Fowler winding-drum plowing engine. The wheels were wider and much more sturdy than those on Ransome's farmer's engine.

side of the engine, another series of gears also used the crankshaft to turn the rear wheels.

The fact that Fowler's engine was self-propelled saved a great deal of time in the field and eliminated the need for horses and hired hands. His attention to detail and precise engineering made the engine efficient and reliable.

The Fowler winding-drum engine and balance plow could turn eight acres of soil a day. The flywheel could operate threshing machines and other farm equipment. Finally, when the work was all done, the machine could haul the same number of wagons as fifteen horses. As word of the unit's capability spread, sales began to rise. One farmer in the Thames Valley, Mr. E. Rack, was so impressed that he sold all fifty-six of his plowing oxen and replaced them with a single Fowler engine and plow.

If the engine had a drawback, it was one that plagued all builders of steam engines: its weight. When the driver took the machine—which weighed six and a half tons—onto soft, wet dirt, the five-foot-tall metal wheels often got stuck. This must have frustrated Fowler a great deal. He probably knew that the most efficient way to plow a field would be to haul the plow directly behind the engine. This arrangement would do away with the need to move the engine, anchor and plow every two rows.

We'll never know if John Fowler was thinking of making a smaller engine for direct plowing. In December 1864, at the age of thirty-eight, he died in a hunting accident. Yet despite his short life, Fowler had invented the first major advance in plowing since the horse-drawn plow had been created 1,000 years before. And as a testament to their usefulness and durability, some Fowler plowing units were still operating as late as the 1930s.

5 Experiments and Improvements

While John Fowler was putting together his plowing unit, other designers in England worked to improve the steam engine itself. None of these machines were called tractors, though they did the work of one. They were known as steam traction engines, steamers, road locomotives or, sometimes, simply engines. One of the most daring was Thomas Rickett's 1858 plowing engine.

Rickett envisioned a self-propelled steamer that could be driven directly across a field towing a plow behind it. This would save a great deal of time by eliminating the need to move the engine, anchor and plow.

Rickett put a digging shaft seven feet long on the back of his machine. The shaft was turned by a heavy chain connected directly to the flywheel crankshaft. As the machine crawled forward, the digging shaft spun around and around, chewing up the hard earth. Chains on the rear of the machine let Rickett haul a second plow behind the digging shaft to smooth out the soil and make it ready for planting.

Rickett's machine was slow-moving and could cover only six acres a day—not quite as good as a Fowler unit. And when fully loaded with water and fuel, it weighed almost ten tons. Owners of a Rickett often found themselves bogged down in soft dirt, especially in the

The Rickett steam engine
with digging shaft

spring, when the ground was wet. Word of the problem spread, and sales of the Rickett engine dropped off.

The massive weight of steamers was their biggest drawback for many years. In order to build up enough steam to pull heavy loads or haul a balance plow across a field, the boilers had to be very large. But the larger the boiler, the heavier was the machine. All this weight was concentrated at the points where the wheels touched the ground, and the intense pressure pushed the wheels into the earth.

Eventually, the metal wheels were made wider, some of them as much as seven feet across! These wide wheels distributed the machine's weight over a greater area and prevented them from sinking into soft dirt. But until wide wheels were a regular feature, most farmers preferred to stick with the engine-cable plowing units.

Many experiments like Rickett's failed. But a number were either fully or partially successful.

For instance, Thomas Aveling introduced "pilot steerage" around 1860. Up until then, the steering for all steamers had been clumsy and imprecise. The engineer had to lean out around the boiler and smokestack to see the road ahead. Often a gust of wind pushed smoke, steam or sparks into his face. Since steering systems re-sponded very slowly, the engineer could find himself in a ditch before he could steer away from trouble.

In Aveling's pilot steerage, a man sat in front, his legs dangling between a wooden frame connected to the front axle. From his vantage point, the pilot had a clear view of the road, free of sparks or other obstructions. At a bend in the road, the pilot simply turned his small front wheel with a rudder, which made the engine's two front wheels follow. Some designers liked Aveling's idea but didn't want to add the steering apparatus to the machine. Their solution was simple. They built a small platform for the pilot at the nose of the machine and moved the steering wheel up.

Others tried to smooth out the ride. The roads and fields were often deeply rutted or full of rocks. If the wheels of a big steamer entered a hole, the entire machine pitched to the side, much like a ship rolling in high waves. The engineer and pilot were thrown around violently, and because the engineer didn't have a steering wheel to hold on to, a number of them tumbled off their machines!

James Boydell thought he had the problem solved in 1864. He saw that metal tracks made railroad traveling smooth no matter what the terrain was like. So Boydell fitted six shoes on each of a steam engine's wheels.

A steamer with Aveling pilot steerage.
The two-wheeled coal tender allowed the
machine to run all day without stopping.

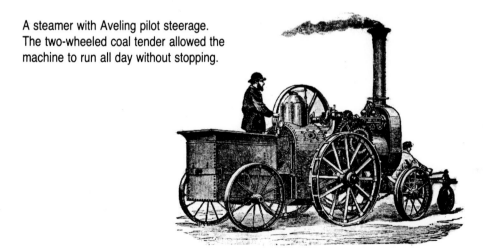

The shoes were pieces of wood, four feet long, faced with iron. They were hinged to the wheel so that they could move freely. As the wheel turned, the shoes swung down to meet the road one after the other. The wheels never actually touched the road surface. They stayed on top of the shoes.

Boydell called his wheel shoes the "endless railroad." He wanted everyone to know that his contrivance provided a smooth, comfortable ride at all times.

It must have been pretty startling to see a ten-ton steamer laboring along, its shoes clanking to meet the ground, but by all reports the device did work. Because the machine's weight was distributed along the four feet of the shoes, the weight wasn't concentrated at the point where the wheels touched the road, and so the machine didn't sink into the soft dirt. The endless railroad created a moving bridge for the machine to ride across small ruts. The shoes also acted as ramps, allowing the steamer to ease itself into and out of larger holes.

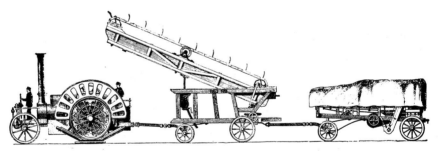

A steam engine fitted with Boydell wheel shoes and front steerage. The machine is pulling a wheat thresher (covered) and hay stacker.

Unfortunately, Boydell's invention had a flaw. On hard, flat roads, the shoes slapped down with so much force they tended to crack. But although Boydell's endless-railway shoes failed, other inventors saw the possibilities in his idea and adapted it to devise a segmented wheel.

The Advance chugs along on Thomson road tires.

R. W. Thomson, of Edinburgh, Scotland, patented his road tires in 1867 and fitted them to the steamer Advance. They consisted of rectangular pieces of rubber mounted on wrought-iron plates. The plates were connected to one another by metal bands.

The Thomson rubber tires absorbed many of the smaller bumps encountered on nineteenth-century roads. The bigger bumps were smoothed somewhat by large metal springs on the front wheel. The Thomson tires worked very well on hard, dry roads. On soft dirt or wet grass, however, the slick rubber slipped badly. Later, thicker and stronger rubber treads were introduced to provide better traction. Eventually, the rubber was replaced by metal for use in tanks and caterpillar tractors.

The improvements made by the Ransome company, John Fowler, Thomas Aveling and R. W. Thomson made steam engines more useful, more reliable and more comfortable to operate. Even so, farmers didn't rush to buy them.

Many still feared steam engines. Those that saw the value of steam were put off by an engine's cost. The average farmer in England earned well under $200 from his small farm. A steamer cost between $1,500 and $2,500, depending on its size. An entire plowing unit went for $4,000.

Since most farmers couldn't afford the machines, builders looked for other ways to use the powerful engines. For instance, William Stephens fitted a fourteen-ton steamer with a derrick in 1862. The derrick could be used to haul heavy objects, such as a railroad locomotive or another steamer.

Thomas Aveling created the first steamroller in 1865. The two front wheels were seven feet tall and two feet wide, while the rear wheel was seven feet tall and five feet wide. Since this engine weighed over thirty tons, it flattened out any road it rolled over.

The first road derrick. The small metal nose in the front was there in case someone needed a push.

The Aveling steamroller rumbles down a road.

In the end, most steamers were used to haul things along public roads, from wagonloads of people, to mail, to cannons and other machinery. They were strong enough to carry forty-ton blocks of granite, or even a house.

The most popular use of steamers was for pulling circuses from town to town. These engines were brightly painted, with large canopy roofs and wrought-iron trimming. Thousands of people lined the road to watch the steamer pull eight or nine wagons loaded with animals and performers, plus the bandwagon. These processions usually drew more people than the circus itself, and some owners actually began featuring the steamer as their main attraction. In time, steamers used to haul circuses became known as "Showman Engines."

An 1890s Showman Engine. The hose on the side made filling the boiler with water easy. The metal bands on the rear wheels, called angle grousers, provided traction.

By the 1860s, steam engines were capable of doing almost any kind of work that required pulling, yet they never became widely used. England's powerful railroad companies, fearing that steamers would take away some of their hauling business, managed to get road tolls for steam engines raised. Then in 1865, they persuaded Parliament to pass the "Red Flag Act." This law forbade steamers from going faster than four miles per hour. It also required that a man walk sixty yards in front of the steamer, waving a red flag to warn pedestrians of its approach. With speeds reduced so much, businesses found it faster and cheaper to ship their goods by railroad or to use horse-drawn wagons.

Steamers were still made and operated, of course. But they never reached their full potential in England. Here the story shifts to the United States, where the vast, untilled regions of the West provided the perfect opportunity for the development and use of steamers.

6 The Case No. 1

Jerome Increase Case left his Oswego, New York, home in 1842 and headed west. He was only twenty-three years old, but he knew exactly where he was going and what he'd do once he got there. His destination was Rochester, Wisconsin, the heart of the new wheat-growing country, and he meant to be a thresherman.

Along with his personal belongings, Case brought six groundhog threshers he had purchased on credit. The groundhog did away with the back-breaking job of removing the kernels of grain from the wheat plants by hand. A treadmill turned by a horse powered the machine. The machine was completely new, but Case was already a master at operating and repairing it.

Case managed to time his trip perfectly. In 1838, the area in which he later settled had shipped a measly 78 bushels of grain. The summer Case arrived, the crop had jumped to 586,907 bushels, and new fields were being cleared and planted every day.

Case sold five of the groundhogs to farmers along the trail. With the remaining machine, he went from farm to farm doing custom threshing for a fee.

Young Case made enough money that year to pay off all his debts and still have plenty to see himself through the long winter. But he didn't sit around waiting for another season of threshing to arrive. He spent the winter trying to improve the groundhog.

All that the groundhog did was beat the grain out of the heads and dump both grain and straw in a pile together. Case devised a thresher that not only beat out the kernels, but also separated them from the straw. The clean grain was delivered to one place, while the straw was blown into a separate pile. The machine worked so well that Case moved to Racine, Wisconsin, and opened his Racine Threshing Machine Works.

For twenty-six years Case built the best threshers in America. He also made the horse-powered treadmills to run them. But as early as the 1850s, the idea of using steam power was beginning to become popular in America. After watching a Hoard & Bradford steam engine in action in New York, Horace Greeley had declared that "threshing will cease to be a manual and become a mechanical operation . . . and this engine will be running on wheels and driving a scythe before it, or drawing a plow behind it, within five years."

A Case thresher in action. The men at the left feed oats to the thresher. The kernels are separated from the plant and channeled to the two men at the side for bagging. The straw is carried along on a conveyor belt for stacking.

An early Case engine

Greeley's enthusiasm was premature, since steam power did not become commonplace in America until well into the 1880s. But such occasional public notice of the potential of steam power set Case and his engineers to work. In 1869, Case introduced his first steam engine, appropriately named the "Case No. 1."

The Case No. 1 wasn't particularly sophisticated or handsome, especially when compared to the British steamers being produced at the time. But like all Case products, it worked well.

The top-mounted cylinders drove the large flywheel. A complicated arrangement of gears connected the flywheel crankshaft to the rear wheel and made it turn. A shaft through the boiler and connected to the front axle by a chain allowed the driver to steer.

Like the British manufacturers of steamers, Case found it hard to sell his early machines. Farmers still feared the effects of steam. Many insurance companies even threatened to cancel policies if a steamer was allowed on the farm to power a thresher. But the biggest drawback was still the selling price. A Case thresher and horse-powered treadmill cost about $325, which included the price of the horse. A steam thresher unit went for $625 to $2,300. Since most farmers had little land—usually under 150 acres—they couldn't afford the price of steam.

Despite poor sales, Case continued to make improvements in his steamers. A spark arrester was added to the chimney. An ingenious device, called the "fusible plug," sensed when water was too low in the boiler and put out the fire before the machine blew up. Steering systems were improved. Case even produced a steamer that could burn coal or wood or straw, depending on the fuel source available. Each year saw new changes in the machines, most to avoid any possibility of mechanical error. To avoid human error in running his engines, Case issued special warnings. One read: "Pure water without any admixture of whiskey is all that is needed to make steam for the J. I. Case & Co. Engine, a fact engineers will do well to remember."

Yet sales remained sluggish. Case sold 75 steamers in 1876 and 109 a year later. Then a major American banker went broke and the era of steam began.

A giant Case steamer

7 The Wheat Boom

The year 1873 opened with the failure of America's biggest banker, Jay Cooke. Cooke's problem was that he held a lot of worthless Northern Pacific Railroad bonds.

Several executives of the railroad found themselves holding the same bonds. Instead of going under like Cooke, at least two of them, George Cass and George Cheney, made a fast and somewhat illegal deal, exchanging their bonds for Federal-grant lands in North Dakota's Red River Valley.

Cass and Cheney wanted to sell their land to settlers from the East and be done with it. But the railroad's land commissioner, J. B. Powers, had another idea.

Powers urged Cass and Cheney to undertake wheat farming on a gigantic scale. He argued that a large factory could mass-produce goods more cheaply than a small one, and thus make a bigger profit. Using the latest machinery and hundreds of men to operate it, they could do the same with wheat.

To prove his point, Powers brought in Oliver Dalrymple, an expert on wheat growing. In the spring of 1875, Dalrymple planted an experimental 1,280 acres. When he brought in an amazing harvest of 32,000 bushels, Cass and Cheney were convinced. By the mid-1880s, the Cass-Cheney-Dalrymple farms totaled 75,000 acres and averaged over 600,000 bushels of wheat a year.

Reporters swarmed over the valley after the second year. Readers heard how 1,500 hired hands worked the farm with horse- and steam-powered machines. But it was the steam engines that captured readers' imaginations.

Stories told of monstrous steamers, their engines throbbing and smoke filling the sky, plowing endless rows or cutting and separating mountains of wheat. Foremen galloped from crew to crew making sure that the engines were running smoothly and the crews were working at full capacity. Tank wagons followed the steamers with water and fuel; grub wagons brought meals to the men so that not a moment was wasted.

News of the Cass-Cheney-Dalrymple success brought in other big investors, and the wheat boom was on. By 1890 the region had six other farms of between 10,000 and 60,000 acres, and more than 2,000 individuals had over 500 acres under cultivation.

This expansion in wheat growing in the Midwest was matched by a similar development in California. One farmer, Dr. Hugh Glenn, was the world's biggest wheat grower in the 1880s, harvesting over 1,000,000 bushels annually.

A couple of enormous Best engines rumble along the San Joaquin Valley hauling combines.

A 110-horsepower Best. The hose running across the front was used for taking on water. The metal ridge around the center of the front wheel is called a guide band. It stopped the wheel from sliding while the machine was making a turn.

During plowing time, Glenn's rigs set off at daybreak, cutting straight furrows for the entire morning. They would halt at noon for lunch, then rumble off again, never stopping until night came. When the single line was finished, the steamer would be turned around and made ready for the next day's plowing.

With the growth of these giant farms came increased traction engine sales. And with sales came new manufacturers of steamers. Hundreds of steamer shops, big and small, opened throughout the country, producing a wide variety of machines. Since patent restrictions weren't enforced very well back then, most manufacturers simply "borrowed" their design ideas from existing machines. Quality varied, too, depending on the skill of the makers. A few factories, however, produced well-made and efficient machines.

One such outfit was the Best Manufacturing Company of San Leandro, California. Best made a giant three-wheeler with the pulling power of 110 horses. The rear wheels were made of iron and were over seven feet tall. Since it was impossible to see the smaller front wheel, an arrow was placed on top of it so the driver knew which way it was pointed.

The Best machines were simple in design but reliable. Many were used by farmers in the San Joaquin Valley. However, for every company that produced a good engine, there were ten that made poorly designed ones.

To prove their steamers were superior, the better manufacturers had to think up new sales techniques. One Case salesman was noted for assembling a large crowd of potential buyers around his machine. Then he'd drive the steamer up a gigantic seesaw and balance it in the middle. Next, he would demonstrate the engine's smooth handling by moving it forward and backward until the seesaw was rocking gently.

Even bigger attention-getters were the "Specials" put together by larger manufacturers such as Case. A train of twenty or more flatcars would be assembled, each beautifully painted and loaded with a steam engine and thresher. The train traveled throughout the wheat country, stopping wherever a crowd of farmers could be gathered. The steamers were fired up and the threshers set into wild motion, while a band or calliope played accompanying music. One company even used its hay-stacking devices to spew advertising handbills out over the crowd.

The most convincing way to demonstrate an engine's power and performance was to enter it in a plowing contest. Hundreds of such

A Case steamer shown in the 1911 Winnipeg plowing contest. Case took home a gold medal.

Painting of a threshing crew in action

contests were held throughout Canada and the United States. The object was to see which engine could plow the most in one day. A reporter at a 1911 contest in Winnipeg, Canada, described the scene as one of "clouds of smoke and hissing steam; a broad prairie stretching for miles without break . . . throngs of eager spectators; imagine all this, add to it the sight of a score of monster engines pulling leviathan plows, and you have a faint picture of the Winnipeg plowing contest. . . ."

Such aggressive promotion did boost the sales of well-made steamers. J. I. Case estimates it sold nearly 36,000 engines between 1880 and 1920.

But although machines were being produced and sold, most of them were to be found only on the larger farms. The millions of farmers who had only a small-acreage farm still couldn't afford the purchase price of an engine. This increased the need for independent custom-plowing and -threshing teams.

An individual would buy a complete plowing and threshing rig. Then he'd take it and a crew of men from farm to farm, plowing in the spring and threshing in the fall for a fee. The pay must have been good. At the opening of the twentieth century there were about 70,000 independent crews throughout the country.

Running one of these outfits required at least six men. One man— usually the owner of the steamer—drove it, while a second kept the fire going and watched the water level. Three or four men were required to operate the thirty-foot-wide plows and other equipment. Another man followed the steamer with a water tank and additional fuel. During threshing season, additional hands were hired to feed wheat into the thresher and stack the straw.

In nonfarming regions, the independent outfits found other ways to use their steamers. A steam engine's flywheel could drive a saw-mill in northern logging territory. Then a train of wagons loaded with fresh-cut wood could be hauled to the nearest train station for shipping.

The work done by this steamer in northern California logging country helped support a number of families.

The 1892 Froehlich gas engine

Even as the era of steam power was in full swing, a new kind of engine appeared—the internal-combustion engine. The power to run a steamer was produced outside the cylinder. Steam from the boiler was fed into the cylinder to drive the pistons and rods. The internal-combustion engine was different. Gas vapor was sprayed directly into the cylinder and then exploded by an electric spark. The explosion *inside* the cylinder pushed the piston and rods.

The earliest internal-combustion engines were extremely crude. John Froehlich built the one shown here in 1892. Not much is known about the Froehlich Gas Engine other than that it could go backward and forward and managed to thresh 62,000 bushels of wheat one year. What we do know is that the Froehlich and most other early internal-combustion engines were noisy and cantankerous machines. Breakdowns were frequent. And if too much gas was exploded in the cylinder, the entire engine often blew sky-high.

Some manufacturers backed away from the new engines, considering them too much trouble compared with steamers. Nevertheless, a few engineers persisted, working to perfect the internal-combustion engine just as men had worked earlier to perfect the steam engine.

Charles Hart and Charles Parr of Iowa spent two years tinkering with their internal-combustion engine, until they got it to run well enough to offer it for sale. In 1903 Hart-Parr sold fifteen gas engines. Of these, at least five were still operating more than twenty-five years later.

Hart-Parr did more than just build a good machine. In order to distinguish their gas engine from the steamers, they came up with a new name for it; they called it a "tractor."

Steam traction engines would still dominate the farm scene for another twenty years. But right on their heels and gaining fast was the gas tractor.

The 1902 Hart-Parr tractor

8 Perfecting the Gas Tractor

If the nineteenth century closed to the powerful throb of steam, the twentieth century opened with the chug-a-chug of the gas-powered tractor. But for the most part, these early machines were erratic at best.

The machines were hard to start, especially during chilly fall mornings. To be certain that the tractor would be ready to work each day during harvest time, some farmers hired boys to keep it running all night.

Since mufflers hadn't been perfected, a gas tractor was much noisier than a steamer. And broken parts were hard to replace. It wasn't unusual for a farmer to have to manufacture his own spare parts if he expected to bring in his crops. One man forced to do this reported paying out $1,500 in one season just to keep his tractor going.

Despite these problems, the advantages of the gas tractor kept their manufacturers confident that farmers would buy them eventually. The fuel used for the engine—kerosene or gasoline—was very inexpensive back then, costing only pennies per gallon. And when a gas engine worked, it produced a great deal of power even though the engine was small. The 1912 Case 20/40 had the power of a 40-horsepower steamer.

A 1912 20/40 Case tractor working a field

The 1920 kerosene-burning Waterloo Boy

More important, one man could operate the machine with ease. A flip of the switch started the engine. The driver checked the gauges himself while the tractor moved along. Special extension arms and ropes permitted him to operate plows and other equipment from his seat.

As the engines were improved and made even smaller, the shape of the tractor changed. Subsequent models looked less and less like the big, bulky steamers they were replacing. Case's 1916 10/20 was considered a neat, compact machine for its time and was capable of generating 20 horsepower. The engine was small enough to be concealed under a metal hood.

The sleek 1916 10/20 Case tractor had a pulling power of 20 horsepower. It also had an arrow over the front wheel to help the driver steer correctly.

Steamers still held the edge in numbers, mainly because farmers were used to them. But by 1917, there were already 14,000 gas tractors in use in the United States. Then Henry Ford entered into the production of tractors and ended steam's dominance forever.

Ford was already rich and famous for his Model T Ford, the automobile he introduced in 1908. It was a squat little car, both dependable and fast, and it was capable of hitting forty miles per hour on a straightaway. Ford kept the design of the engine simple for easy maintenance, and he made sure parts were readily available. The Model T's biggest selling point, however, was its price.

When it first came out, the Model T sold at a reasonable $850. Then the price actually went down for the next sixteen years! In 1924, a Model T cost just $290. These cars were so good that some farmers even converted them for use as tractors.

Ford's idea for a tractor was just as basic. He wanted to mass-produce a small but powerful machine that almost every farmer could afford. The Fordson tractor rolled off the assembly line in 1917.

The Fordson was a success from the start. Its sales were boosted because of the effects of World War I, raging in Europe since 1914. The war had destroyed millions of acres of farmland in Europe and forced up the price of wheat. American farmers rushed to grow as much as possible. When the United States entered the war in 1917, the result was a labor shortage, as hundreds of thousands of men were drafted into the army.

This 1917 Fordson probably did a good deal of work in its time. The winch and cable on the front were added for pulling up tree roots.

Farmers who had refused to trade in their steamers for gas tractors were practically forced to do so during the war. They simply couldn't find enough men to help keep the big steamers going. During the four years of World War I, over 71,000 gas tractors were sold. And the year following the 1918 Armistice, another 73,000 machines were produced.

Of course, many farmers stuck with steam. A number of steam-powered machines were still running even as late as the 1950s. But fewer and fewer were being made. Case, the world's largest steam-engine manufacturer, made its last steamer in 1926.

9 The Tractor Comes of Age

The Fordson set the pattern for tractor design after World War I. Soon other automobile makers such as General Motors and Willys began producing similar machines. John Deere and Company, International Harvester and other manufacturers of farm tools saw that the market for gas-engine tractors was growing rapidly and entered the field. Even well-established tractor makers like Case felt the pressure to modify their tractor designs—and costs.

Many of the improvements in these tractors were borrowed from those being used in automobiles. To help engines warm up in cold weather, engine exhaust was diverted around fuel-air passages. The warm vapor ignited more easily in the cylinder, and the farmer could drive off to work sooner. Mufflers were improved to quiet the rugged engines. Once these tractors proved popular with farmers, the companies began searching for ways to attract even more buyers to their particular machines.

John Deere came out with its famous GP model in 1928. "GP" stands for "general purpose," and this machine came with four separate power sources, so that it could be put to a variety of uses. It could pull plows and seeders. It had a belt pulley similar to the steamer's flywheel to operate haying devices and threshers. It could be hooked up to small mowers or combines. And it had a power lift that allowed the driver to raise and lower equipment with ease. No matter what kind of equipment the farmer might already have in his barn, the GP could be used to run it.

A few years later, Deere began selling the GP Orchard model. The deep-skirt fenders covered the rear wheels down to the hubs, as well as the engine. This allowed the driver to slide his tractor between low-growing branches without injuring the trees or getting them caught in the spokes or engine.

A Case 10/18 tractor of the same era could be equipped with steering extension. The driver sat back on his wagon as he drove to market. From this perch, the driver could make sure everything

A post–World War I Case tractor

A 1931 John Deere GP. The five-inch-long cleats on the rear wheels are called spade lugs. They dug into the ground for better traction.

A Case tractor with steering extension. Note the solid rubber tires on front and rear wheels.

was okay with his wagons. He could also operate the wagon brakes. Two ropes let him make the tractor go faster and stop. Another attachment turned the Case into a snowplow.

The Case tractor could be fitted with another piece of special equipment—rubber tires. As more and more dirt roads were paved over in the 1920s and 1930s, farmers found it necessary to cover their metal wheels. The spade lugs, or cleats, bit into the road surfaces the same way they bit into the soil, so many localities prohibited the use of metal wheels on paved roads to avoid damage.

In the early 1920s rubber tires were fitted over the metal ones the way shoes go over your feet. Gradually, solid rubber tires were produced. When the farmer went to market, he exchanged the metal wheels for rubber ones. Air-filled or pneumatic tires became standard equipment in the 1940s, but many farmers preferred metal wheels, and these continued to be available through the 1950s.

It seemed that whenever a group asked for a modification on a tractor for their special line of work, manufacturers tried to comply. When loggers in the Northwest demanded a machine that could maneuver through mud and snow and hilly terrain, the caterpillar tractor was designed. Its long metal treads distributed the weight of the machine along its entire length. Since the weight wasn't concentrated in any one area, the machine didn't sink into the soft dirt. Eventually, farmers in wet or hilly areas found that the metal bands of a caterpillar had better traction than regular tractor tires, and they began using them, too.

Other farmers complained that the standard placement of the wheels wasn't right for their particular crop rows. The broad leaves of cabbage might require a wider row than, say, corn. Deere re-

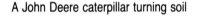

A John Deere caterpillar turning soil

John Deere's Model A with air-filled tires

sponded with the Model A in 1934. The chief feature of the A was that the rear wheels could be moved in or out on the axle, so that the space between them could be adjusted as needed, depending on the width of the row.

A second feature of the A was its new design. The square radiator of the older models was made smaller and the hood tapered. This gave the driver better visibility while moving through narrow crop rows. The farmer could also order a machine with two center-mounted front wheels. This made traveling up the crop rows easier. All the driver did was aim the center of the hood up the middle of the row and the wheels would follow, fitting into the two side strips easily.

What finally emerged was the kind of tractor most of us associate with farm work today. And as the tractor changed, so did the business of manufacturing it. Engineering qualities of a gas engine had to be extremely precise. A slight flaw not only resulted in a poorly performing engine, but it also could cost a manufacturer millions of dollars in repair bills. And in order to compete in price, machines had to be mass-produced by the thousands. To meet these requirements an enormous amount of money was needed, and only a large corporation had such funds.

So as the gas tractor set a new generation of farm machinery into motion, it also did away with the individual entrepreneur. It would no longer be possible for a John Fowler or J. I. Case to take an idea for a machine and set up manufacturing on a small scale.

10 New Markets, New Machines

The dependable internal-combustion engine did away with the steamer and changed the look of the tractor. It also drove out the small, independent manufacturer. But oddly enough, it didn't replace the oldest form of power on the farm right away. As late as the mid-1930s, there were still more than 24,000,000 horses used to haul and operate farm equipment.

A lot of farmers had tractors to plow their fields, but preferred horses for the meticulous work of cultivation, such as pulling fertilizing and weeding devices. Others didn't think their small farms could justify even the cost of a cheap Fordson. Manufacturers saw these hundreds of thousands of farmers as a potential market and set their designers to work to shrink the tractor and shrink its selling price.

One such tractor was John Deere's Model H, released in 1934. The Model H weighed only 2,000 pounds, half what a Model A weighed. Its very narrow, tapered hood and high seat gave the farmer an unimpaired view of where he was going. And at $650, thousands of farmers could afford it.

Like the Fordson before it, the H tractor fit the needs of its intended market perfectly. The H was so popular that John Deere made this model for eighteen years and sold over 250,000 of them. As sales of the H and other small tractors rose, the number of horses

The tiny Model H pulling an automatic seeder

declined. Today there are only about 5,000,000 horses around, and not many of these do any heavy farm labor at all.

At about the same time, tractor makers began looking at another potential market—the homeowner. The twenties and thirties saw the growth of suburbs as middle-income families left the crowded cities for more open country. Flight from the city intensified after World War II.

In addition to breathing fresher air and looking at trees, homeowners also planted gardens. Since most of these gardens were under a half acre in size, homeowners had no need for a large tractor. Instead, these people turned the soil and planted seed by hand. Then the Bolens Company came out with the first riding tractor in 1931.

It was a crude machine from any angle. A tiny one-cylinder gas engine turned the two front wheels. The driver sat behind the plow blades, holding on to the steering handle. When charging around the garden, the driver had to be careful to hang on tightly at all times or risk being thrown off.

Eventually, more sophisticated and comfortable home tractors appeared. These machines may look dinky next to a full-sized farm tractor, but they can operate an amazing variety of equipment. The

The first riding garden
tractor by Bolens

A modern Bolens riding tractor with
lawn-mowing attachment

modern Bolens can pull a plow, dump cart, cultivator, seeder, lawn sweeper, rake or fertilizer. It can even be used to operate a snow-blower or rotary broom. For the homeowner who didn't have a garden but had a large lawn, a riding mower was designed. It is nothing more than a $2^{1}/_{2}$-horsepower tractor with cutting blades attached to its belly.

These changes in tractor design were matched by a changing way of life on the farm. The Great Depression of the 1930s forced thousands of farmers to abandon the land and look for work in factories. During the forties and fifties, many farm children left to seek work in the cities and suburbs. As the farm population declined, bigger farms bought out the smaller ones. These giant farms began demanding tractors with much more power.

To fill these power needs, a new form of internal-combustion engine was fitted to tractors in the late forties: the diesel. Unlike the gas engine, the diesel doesn't use an electric spark to ignite the fuel in the cylinder. Instead, extremely hot compressed air ignites the fuel. This produces a cleaner, bigger explosion and more power.

Diesel engines aren't as large as the older gas engines, either. The John Deere 4020 diesel can produce more than 91 horsepower; in comparison, the older model GP managed to generate only about $15^{1}/_{2}$ horsepower. Diesel engines are also simpler and less likely to break down, and diesel fuel is cheaper than gasoline.

But the biggest advantage of a diesel is the engine's ability to work. Tractors can now plow up to 125 acres in a single day. And the added power means that the tractor can do several jobs at once. For instance, a 4020 can haul a disk harrow and planter at the same time. During haying time, it can pull a hay baler, a bale ejector and the trailing wagon to carry the newly made bales.

In addition to power, the diesels provide more comfort and safety. Padded seats have replaced the old hard-metal ones. Power steering makes maneuvering easier. Antiroll devices have made the tractor safer in hilly areas.

Giant tractors that rivaled the old steamers began appearing. The John Deere 8010 was eight feet tall and almost twenty feet long. It weighed 19,700 pounds and could pull a plow thirty feet wide.

The trend toward ever-greater pulling power continues today. In the 1970s turbocharging was added to the engines. A turbocharger allows even more fuel to be exploded in the cylinder, generating more power with each stroke of the piston.

A big diesel-powered John Deere tractor hauling an automatic plow and seeder

The driver of this 8010 is sitting eight feet above the ground. Note the stairs at the side for getting onto the machine.

Additional traction is gotten from "dualing"—that is, adding an extra set of tires to the rear axle. Ballast such as sand or water can be put in the six-foot-tall tires to help them dig in even more.

The newest tractors are four-wheel drive, turbocharged machines that can develop over 225 horsepower. Inside, the driver rides in air-conditioned comfort while listening to his stereo tape deck.

A new John Deere tractor plows in hilly terrain.

11 What the Future Holds for Tractors

The history of tractor development has been one of extremely slow change. At first, this was caused by farmers' reluctance to switch from reliable horsepower to the newfangled steam machines. Nowadays, farmers welcome any sort of mechanical improvement that will make farming a little easier. Unfortunately, the high cost of retooling a tractor assembly line makes any changes difficult and expensive. In fact, a new tractor model is usually designed to remain on the market for at least ten years.

Even so, there have been some modest attempts to improve tractors in recent years. One dragstrip racer mounted an F-14 jet engine on a giant tractor. When fired up, his machine generates around 150,000 horsepower. So much power makes the tractor hard to handle, very expensive to run and difficult to repair. It is useless for day-to-day farm work, but its power is a big asset in pulling contests held at state agricultural fairs!

Another individual has placed helicopter blades under a tractor. This allows his machine to float *above* the crops and eliminates any damage that might be caused by massive tires. The problem with his machine is that it has almost no pulling power.

Such space-age tractors are fun to think about, but since they have no practical use on a farm, they'll remain curiosities. The truth is that the tractors being used on farms right now *are* the tractors of the future. Their turbocharging units might be modified to boost horsepower and other small changes may increase the number of jobs the tractor can handle. But, essentially, the tractors that appear five and ten years from now will look almost exactly like those that cut the long furrows all over the world today.

An eight-wheel John Deere 8430 rumbles over a hill.

For Further Reading

Alexander, E. P. *Iron Horses: American Locomotives, 1829–1900*. New York: W. W. Norton & Company, 1941.

Deere & Company. *John Deere Tractors, 1918–1976*. Moline, Ill.: John Deere & Company, 1976.

Holbrook, Stewart H. *Machines of Plenty: A Chronicle of an Innovator in Construction and Agricultural Equipment*, rev. ed.; updated by Richard G. Charlton. New York: Macmillan Publishing Company, 1976.

Hughes, W. J. *Traction Engines Worth Modelling*. London: Percival Marshall & Company Ltd., 1950.

Norbeck, Jack. *Encyclopedia of American Steam Traction Engines*. Glenn Ellyn, Ill.: Crestline Publishing Company, 1976.

Pursell, Carroll W. *Early Steam Engines in America: A Study in the Migration of a Technology*. Washington, D.C.: Smithsonian Institution Press, 1969.

Stein, Ralph. *The Treasury of the Automobile*. New York: Golden Press/A Ridge Press Book, 1961.

Wendel, Charles H. *Encyclopedia of American Farm Tractors*. Sarasota, Fla.: Crestline Publishing Company, 1979.

Williams, Michael. *Farm Tractors in Color*. New York: Macmillan Publishing Company, 1974.

Index

Page numbers in *italics* refer to illustrations